COMPETITIVE MATHEMATICS 1

INTRODUCTION

This objective mathematics series provides a basic and challenging problem of mathematics from particular topics. It can be used to brush up ones basics and checking up the preparation level of particular topics. It is equally helpful to the traditional classes as well as competitions. It can be also taken as a revision material for any competition which includes the test of basic mathematics. If you want to grasp the subject before practicing these multiple choice questions, you can go through the website http://www.ncert.nic.in/ncerts/textbook/textbook.htm and down load the free copy of mathematics books and after having command on the topic practice it. For revision purpose, important points are given at the starting of each topic.

CONTENTS

1. PROBABILITY

SOME IMPORTANT POINTS

- The empirical probability can be written as P(E)
- Empirical probability P(E)=Number of outcomes/Total number of outcomes
- The probability of an event can be lies between 0 and 1
- 1 is the probability of a sure event
- 0 is the probability of an impossible event
- We can say that not P(E) is complement of P(E)
- We also say that P(E) and not P(E) are complement events
- P(E)+not P(E)=1
- The probability of an event is likely to 0< P(E)<1
- Where P(E) means probability of an event

1. STATISTICS

SOME IMPORTANT POINTS

➢ The data are the collection of facts and figures.

➢ The collected data can be represented graphically in the form of bar graphs, histogr and frequency polygons.

➢ The mid points at the class intervals are called class marks.

➢ The mean is found by adding all observations and divide it by total number of observations. X = Sum of total observations/Total number of observations.

➢ Where x mean of observation.

➢ The symbol "$\sum$" we mean (sigma).

➢ The word "n" we use means total number of observation.

➢ The median is found by finding $(n+1/2)^{th}$ observation, when n is an odd number.

➢ The median is found by finding $(n/2)^{th}$ and $(n/2+1)^{th}$ observation, when n is an even number.

➢ The mode is that observation which occurs frequently.

1. PROBABILITY AND STATISTICS

1. Two coins are tossed simultaneously 200 times and we get.

    Two head: 110 times

    One head: 50 times

    No head: 40 times

    What is the probability of two heads?

    (a) 50/200   (b) 40/200

    (c) 100/200   (d) 110/200

2. The heights (in cm) of 9 students of a class are

    155, 160, 145, 149, 150, 147, 152, 144, 148

    What will be the mean of this data?

    (a) 140   (b) 200

    (c) 150   (d) 100

3. The points scored by a Kabbadi team in a series of matches are?

    17, 2, 7, 27, 15, 5, 14, 8, 10, 24, 48, 10, 8, 7, 18, 28

    What will be the median of this data?

    (a)   (b) 10

    (c) 8   (d) 20

4. 20 students get following marks (out of 10) are

    4, 6, 5, 9, 3, 2, 7, 7, 6, 5, 4, 9, 10, 10, 3, 4, 7, 6, 9, 9

What will be the mode of this data?

(a) 1   (b) 4

(c) 9   (d) 6

5. If the median of this data is 63.

29, 32, 48, 50, x, x+2, 72, 78, 84, 95. What is the value of x?

(a) 50   (b) 61

(c) 42   (d) 60

6. In a cricket match a batsman hits boundary 4 times out of 24 balls he plays. What out probability that he did not hit boundary.

(a) 16   (b) 18

(c) 20   (d) 14

7. The weight (in kg) of 10 students 15, 20, 25, 35, 38

40, 45, 32 and 28.

What is the mean of this data?

(a) 3.5   (b) 3.8

(c) 2.6   (d) 2.8

8. 2 coins are tossed simultaneously 10 time and we get:-

Two heads: 4 times

One heads: 2 times

No head: 1 time

Find the probability of one head:-

(a) 4/10   (b) 2/10

(c) 3/10   (d)0/10

9.   In class 6th students obtained marks (out of 20) are:-

5, 10, 6, 7, 9, 12, 13, 16, 19, 18, 17, 15, 14, 12, 8.

What will be the mean of this data?

(a) 10.45   (b) 14.32

(c) 12.710   (d) 8.22

10.   What will be the median of this data:-

5, 10, 6, 7, 9, 12, 13, 16, 19, 18,

(a) 9   (b) 12

(c) 7   (d)13

11.   What will be the mean of the following data?

5, 10, 6, 7, 9, 12, 13, 16, 19, 18,

(a) 12.2   (b) 10.5

(c) 11.5   (d) 8.4

12.   In class 6th students obtained marks (out of 20) are:-

5, 10, 6, 7, 9, 12, 13, 16, 19, 18, 17, 15, 14, 12, 8.

What will be the mode:-

(a) 10   (b) 19

(c) 18   (d) 12

13. The following number of goals were scared by a team in a series of 10 matches are.

2, 3, 4, 0, 5, 1, 3, 3, 4, 3

13. (1)   what will be the mean

    (a) 2.6   (b) 3.2

    (c) 2.9   (d) 4.1

13. (2)   what will be the mode

    (a) 2   (b) 4

    (c) 3   (d) 5

13. (3)   what will be the median

    (a) 5   (b) 1

    (c) 3   (d) 4

14. What will be the mean of the data?

29, 32, 48, 50,72, 78, 84, 95.

(a) 61   (b) 57.8

(c) 50   (d) 55.2

15. The range of this data:-

25, 30, 38, 20, 36, 40, 45, 15

(a) 20   (b) 30

(c) 13   (d) 15

16. 80 bulbs are defected out of 200 bulbs what is the probability of good bulbs.

(a) 80/200   (b) 180/200

(c) 100/200   (d) none

17.   Mode of the data.

15, 14, 19, 12, 20, 25, 38, 40, 12, 15, 15, 14, 13, 12, 12, 38

(a) 15   (b) 38

(c) 14   (d) 12

18.   The median of the data:-

78, 56, 22, 45, 36, 93, 68

(a) 45   (b) 36

(c) 22   (d) 56

19.   Two coin are tossed 1000 times and the outcomes are:-

No. of head	2	1	0
Frequency	200	250	550

Based on this data the probability for 1 head is

(a) 1/5   (b) 1/4

(c) 4/5   (d)3/4

20.   550 bulb are packed in a carton 28 bulbs are defective. Find the probability of not defective bulb.

(a) 522/550   (b) 28/550

(c) 0/550   (d) none

21.   Range of the data:-

25, 38, 46, 39, 58, 32, 93, 16

(a) 70   (b) 6.6

(c) 77   (d) 64

22. In a medical examination of students of a class the blood groups are:-

Blood group	A	AB	B	O
No. of students	10	13	12	5

A student is selected at random from the class. The probability that he/she has blood group of O:-

(a) 5/40   (b) 13/40

(c) 3/10   (d) 1/8

23. A coin is tossed one time find the probability of getting tails?

(a) 2/2   (b) 1/2

(c) 0/2   (d) 3/2

24. 52 cards are in a pack, and then what is the probability of getting a face card?

(a) 4/13   (b) 3/13

(c) 5/13   (d) 1/13

25. A dice is tossed one time what is the probability of getting a prime number.

(a) 0/2   (b) 1/3

(c) 1/2   (d) 1/7

26. We toss a dice onetime then what is the probability of getting a number less than 5.

    (a) 2/3   (b) 1/3

    (c) 0/3   (d) 24/30

27. In a cricket match a batman play 30 balls and he did not hits boundary 8 times then what is the probability of hitting boundary.

    (a) 22/30   (b) 8/30

    (c) 26/30   (d) 24/30

28. There are 3 red balls, 4 black balls and 5 white balls in a bag. One ball is picked up randomly then what is the probability of getting a black ball.

    (a)1/2   (b) 1/4

    (c) 1/3   (d)1/1

29. 500 times we tossed a coin with following frequencies:-

    Head-255 and tail-245 what is the probability of getting tail?

    (a) 60/100   (b) 30/100

    (c) 49/100   (d) 80/100

30. In 250 consecutive days weather forecasts 180 times are correct. Find the probability of getting not correct.

    (a) 40/250   (b) 50/250

    (c) 20/250   (d) 70/250

31. There are 36 student are in 9th class. 20 students are boys. Find the probability of girls in the class.

    (a) 10/18   (b) 5/18

(c) 8/18   (d) 3/18

32.   5 to 15 number, find the probability of having an odd number.

       (a) 1/6   (b) 2/18

       (c) 8/8   (d) 4/9

33.   814 families have 2 children out of 1500 families and rest of families has 1 child. Find the probability of families having 1 child.

       (a) 500/700   (b) 300/700

       (c) 343/750   (d) 100/700

34.   In word 'INDIA' what is the probability of getting letter I?

       (a) 2/5   (b) 3/5

       (c)1/5   (d) 0/5

35.   56 apples are in a bag, 19 were rotten. One apple is chosen at random; find the probability of getting a fresh apple?

       (a) 38/56   (b) 30/56

       (c) 37/56   (d) 32/56

36.   A dice is tossed one time, what is the probability of getting number 7:-

       (a) 3/8   (b) 1/3

       (c) 0/7   (d) 2/6

37.   These are 3 green card, 5 white and 6 black cards. One card is drawn at random. What is the probability of having a card which is not green?

       (a) 11/14   (b) 12/15

       (c) 16/15   (d) 1/21

38. There 50 student are in a class, 70% were passed. What is the probability of failing the students?

   (a) 0/50   (b) 9/11

   (c) 3/10   (d) 12/15

39. 52 cards are in a back, what is the probability of getting a red king?

   (a) 3/20   (b) 0/20

   (c) 1/26   (d) 28/28

40. 67 vehicles are in a locality, 39 are black. What is the probability of a vehicle which is not black?

   (a) 39/67   (b) 0/67

   (c) 28/67   (d) 67/67

41. In a survey 250 student were conducted about the subject statistics, in which 143 student like statistics. Find the probability of student who does not like mathematics?

   (a) 210/250   (b) 200/250

   (c) 107/250   (d) 110/250

42. The word 'CLASSES' what is the probability of getting letter's'?

   (a) 2/7   (b) 1/7

   (c) 3/7   (d) 0/7

43. There are 47 students are in a class, 29 student studies home science and rest student studies drawing. Find the probability of drawing student.

   (a) 18/47   (b) 29/47

   (c) 16/47   (d) none of these

44. In ward 'MATHEMATICS' what is the probability of a getting a vowel?

(a) 4/11 (b) 3/11

(c) 0/11 (d) 1/11

45. If we consider the number from 1 to 70, what is the probability of number which is divisible by 7?

(a) 9/68 (b) 10/60

(c) 12/58 (d) 0/70

46. In ward 'ESSENTIAL' what is the probability of letter 'E'.

(a) 1/9 (b) 0/9

(c) 2/9 (d) 3/9

47. 52 cards are in a pack. What is the probability of getting spades.

(a) 0/52 (b) 13/52

(c) 1/52 (d) 4/52

48. There are 2 blue balls, 5 white and 3 green balls one ball is picked up randomly what is the probability of getting green ball?

(a) 3/10 (b) 5/10

(c) 2/10 (d) 0/10

49. In ward 'PAINTING' what is the probability of getting a vowel.

(a) 3/8 (b) 2/8

(c) 1/8 (d) 0/8

50. A coin is tossed 500 times with following frequencies

Head -255, Tail-245

What is the probability of getting tail?

(a) 52/110

(b) 70/110

(c) 49/100

(d) 80/110

51. A football team play 10 matches and get 2, 3, 5, 4, 0, 1, 3, 3, 4, 3 no. of goals. Find mean

(a) 2.8

(b) 2.5

(c) 2.4

(d) 2.6

52. Find the median for the above question.

(a) 3

(b) 0

(c) 4

(d) 1

53. Find the mean for:-

4, 3, 7, 0, 0, 6, 8

(a) 4

(b) 3

(c) 7

(d) 10

54. If the mean of 6, 8, 5, 7, x and 4 is 7 then find the value of x.

(a) 8

(b) 10

(c) 6

(d) 12

55. What is the mean of first there natural numbers.

(a) 0

(b) 3

(c) 2

(d) 5

56. One student has scored the marks in the subjects as.

70, 64, 56, 54, 51.What is the mean?

(a) 54   (b) 62

(c) 59   (d) 80

57.   What is the range of 42, 40, 80, 69, 47 and 56?

(a) 60   (b) 38

(c) 40   (d) 12

58.   Find the mode for:-

7, 9, 12, 13, 7, 12, 15, 7, 12, 7, 25, 18, 7

(a) 12   (b) 15

(c) 7   (d) 13

59.   Find the mean for first three whole numbers.

(a) 1   (b) 0

(c) 2   (d) 3

60.   The class marks as

47, 52, 57, 62, 72, 77, 82,

What is the class size?

(a) 23   (b) 20

(c) 5   (d) 10

61.   What is the class mark for the class interval 18-26?

(a) 22   (b) 20

(c) 18   (d) 26

62. 2, 2, 3, 5, 4, 0, 1, 3, 3, 4, 3 find the mode

    (a) 3   (b) 2

    (b) 4   (d) none of these

63. 47, 52, 57, 62, 67, 72, 77, 82, these are class marks find the class limits for the first class marks.

    (a) 44-48   (b) 48-52

    (c) 44.5-49.5   (d) 45-50

64. The mean of 2, 3, 4, 5, 6, and 4 is

    (a) 4   (b) 2

    (c) 3   (d) 0

65. 14, 25, 14, 28, 18, 17, 14, 23, 22, 14, 18 find its mean.

    (a) 18.8   (b) 16.2

    (c) 10.5   (d) 14.4

66. 7, 9, 12, 13, 7, 12, 15, 7, 12, 15, 7, 12, 7, 25, 18, 7 find the median.

    (a)15   (b) 11.6

    (c) 9   (d) 12

67. For the class interval 21-25, what is the upper limit?

    (a) 25   (b) 21

    (c) 0   (d) none of these

68. Find the range

    43, 70, 60, 80, 50

(a) 37 (b) 43

(c) 40 (d) 45

69. For which value of p, the given data has mode 5.

 1, 2, 5, 7, 5, 2, 7, 5, 9, 2, 3, p, 11

(a) 5 (b) 2

(c) 3 (d) 9

70. 1, 2, 5, 7, 5, 2, 7, 5, 9, 2, 3, 11 find the mean

(a) 4.9 (b) 2.6

(c) 4.4 (d) none of these

71. The median of 4, 4, 3, 5, 6, 2, 9 and 8 is

(a) 4 (b) 3

(c) 5 (d) 6

72. 40, 42, 80, 69, 56, 47, find the mean

(a) 48.4 (b) 50.2

(c) 55.6 (d) 42.8

73. If the tally marks of a given data is IIII then what is its frequency?

(a) 9 (b) 8

(c) 4 (d) 0

74. 23, 54, 10, 33, 43, 25, 48 what is the mean?

(a) 28.4 (b) 30.2

(c) 33.7 (d) none of these

75. 7, 9, 12, 13, 7, 12, 15, 7, 12 what is the median?

(a) 12   (b) 9

(c) 13   (d) 15

ANSWERS:

Q.	A.	Q.	A.	Q.	A.	Q.	A.	Q.	A.
1	D	16	B	31	C	46	C	61	A
2	C	17	A	32	D	47	B	62	A
3		18	A	33	C	48	A	63	C
4	C	19	B	34	A	49	A	64	A
5	D	20	A	35	C	50	C	65	A
6	C	21	C	36	C	51	A	66	B
7	B	22	A	37	A	52	A	67	A
8	B	23	B	38	C	53	A	68	A
9	C	24	B	39	C	54	D	69	A
10	B	25	C	40	C	55	C	70	A
11	C	26	A	41	C	56	C	71	A
12	D	27	A	42	C	57	C	72	C
13(1,2,3)	C	28	C	43	A	58	C	73	C

| 14 | A | 29 | C | 44 | A | 59 | A | 74 | C |
| 15 | A | 30 | D | 45 | C | 60 | C | 75 | B |

2. STATISTICS

SOME IMPORTANT POINTS

➤ The mean for grouped data at can be found by different methods.

➤ Direct method $x = \dfrac{\sum fixi}{\sum fi}$

   Where , fi=frequency and xi=class interval

➤ Assumed mean method $=x=a+\dfrac{\sum fidi}{\sum fi}$

   Where we can find di=xi-a

➤ Step deviation method $=x=a+h(\dfrac{\sum fiui}{\sum fi})$

   Where, h=to be chosen appropriately

   ui=xi-a/h

➤ The mode for grouped data can be found by
➤ Mode $=l+(f_1-f_0/2f_1-f_0-f_2)*h$
➤ Where l= lower limit
➤ The commutative frequency of a data is obtained by adding the frequency of all class preceding the given class
➤ The median for grouped data can be found by
➤ Median $=l+(n/2-cf/f)*h$
➤ The empirical relation between mean, mode and median
➤ 3median=2mean+mode

2. STATISTICS

1. The cumulative frequency curves are also called?

    a. Class mark   b. O gives   c. Class mark   d. None of these

2. The Greek letter Σ means?

    a. Summation   b. Frequency   c. Varying   d. None of these

3. The direct method to finding the mean consists?

    a. Deviation   b. Class mark   c. Step deviation   d. None of these

4. The assumed mean method consists?

    a. Deviation   b. Class mark   c. Step deviation   d. None of these

5. The step deviation method consists?

    a. Deviation   b. Class mark   c. Step deviation   d. None of these

6. The mean obtained by all three methods is?

    a. Different in all methods   b. Increases in some method

    c. Same   d. None of these

7. The class size (h) of class interval 10-25 is?

    a. 10   b. 15   c. 25   d. 35

8.   The wood x means?

   a. Mode   b. Median   c. Mean   d. Deviation

9.   The direct mean method has the formulae of?

   a. $\dfrac{\sum fixi}{\sum fi}$ b. $\dfrac{\sum fiui}{\sum fi}$ c. a+ $\dfrac{\sum fidi}{\sum fi}$ d. None of these

10.   The assumed mean method has the formulae of?

   a. $\dfrac{\sum fixi}{fi}$ b. $\dfrac{\sum fiui}{\sum fi}$ c. a+ $\dfrac{\sum fidi}{\sum fi}$ d. None of these

11.   The step deviation method has the formulae of?

   a. $\dfrac{\sum fixi}{\sum fi}$ b. $\dfrac{\sum fiui}{\sum fi}$ c. a+ $\dfrac{\sum fidi}{\sum fi}$ d. None of these

12.   To find (ui) we have?

   a. xi –a   b. xi – a/h   c. fiui   d. None of these

13.   To find (di) we have to put?

   a. xi – a   b. xi – a/h   c. fiui   d. None of these

14.   To find (a) we have to find?

   a. Last term in class mark   b. Middle term of frequency

   c. Middle term of class mark   d. None of these

15.   The class size of class interval 25-30 is?

a. 25   b. 20   c. 75   d. 50

16.   The mode is a value of observation having the?

a. Lowest   b. Maximum frequency   c. Middle frequency   d. None of these

17.   The modal class is a type of class interval with?

a. Class   b. Size   c. Frequency   d. Cumulative

18.   The lower limit of modal class (7-9) is?

a. 9   b. 7   c. 2   d. 16

19.   The class size of modal class (2-4) is?

a. 4   b. 6   c. 2   d. 3

20.   The word (f_0) means frequency of?

a. Class preceding   b. Class succeeding

c. Maximum value   d. None of these

21.   The wood (f_2) means frequency of?

a. Class preceding   b. Class succeeding   c. Maximum value   d. None of these

22.   The word (f_1) means frequency of?

a. Class preceding   b. Class Succeeding   c. Maximum value   d. None of these

23.   The mode of a data has the formulae of?

a. $\dfrac{\sum fixi}{\sum fi}$ b. $l+(f_1- f_0 /2f_1- f_0 - f_2)*h$

c. l+ (n/2- cf/f)*h   d. a+ (n/2+1)*h

24. To finding median of underground data, we arrange data values of the

     observations in?

     a. Ascending order   b. Descending order

     c. Equal frequencies   d. None of these

25. The word (n) means in median?

     a. Maximum Frequency   b. Middle term

     c. Sum of frequencies   d. None of these

26. If (n) is odd, then the median is?

     a. (n+1/2)   b. (n/2+1)   c. (n/2)th observation   d. Both (b) &(c)

27. If (n) even, then the median will be average of?

     a. (n+1/2)th observation   b. (n/2+1)th observation

     c. (n/2)th observation   d. Both (b) &(c)

28. The cumulative frequency distribution of more than type consists?

     a. Lower limits   b. Upper limits   c. Less than type   d. All of
above

29. The cumulative frequency distribution of less than type consists?

     a. Lower limits   b. Upper limits

     c. More than type or equal type   d. All of above

30. The upper limit of class interval (15-30)is?

     a. 15   b. 45   c.30   d. 5

31. The median class is a type of class whose cumulative frequency is greater or

nearest to?

a. n   b. n/2   c. (n+1/2)   d. (n/2+1)

32.   The word (cf) means cumulative frequency of class?

a. Preceding the median class   b. succeeding the median class

c. Equal to median class   d. None of these

33.   The word (f) in median means?

a. Maximum frequency   b. Frequency of median class

c. Middle of frequency   d. None of these

34.   The median of a given data has the formulae of?

$$ \frac{\sum fixi}{\sum fi} $$

a.   b. l+$(f_1 - f_0/2f_1 - f_0 - f_2)$*h

c. l+$(n/2 - cf/f)$*h   d. None of these

35.   The empirical relationship between the three measures is?

a. 3Median = Mode+2Mean   b. 3Mode = Median+2Mean

c. 3Mean = Mode+2Median   d. 3Mode = 2Median+Mean

36.   The term marked at the horizontal axis of bar graph is?

a. Lower limits of class intervals   b. Upper limits of class interval

c. Cumulative frequency   d. class intervals

37.   The term, we mark at the vertical axis of cumulative frequency curve of less

Than type is?

a. Lower limits of class intervals   b. Upper limits of class intervals

c. Cumulative frequency   d. Class intervals

38. The term, we mark at the horizontal axis of a cumulative frequency curve of
    More than type is?

    a. Lower limits of class intervals   b. Upper limits of class intervals

    c. Cumulative frequency   d. Class intervals

39. To find median through o gives without talking any point on cumulative
    frequency, we need?

    a. Less than type o gives   b. More than o gives

    c. Class interval o gives   d. Both (a) &) (b)

40. Which measure is that the most frequently used measure of central
    tendency?

    a. Mean   b. Median   c. Mode   d. None of these

Answers:

Q	A	Q	A	Q	A	Q	A	Q	A
1	B	9	A	17	C	25	C	33	B
2	A	10	C	18	B	26	A	34	C
3	B	11	B	19	C	27	D	35	A
4	A	12	B	20	A	28	A	36	D
5	C	13	A	21	B	29	B	37	C
6	C	14	C	22	C	30	C	38	A
7	B	15	A	23	B	31	B	39	D
8	C	16	B	24	A	32	A	40	A

3. POLYNOMIAL 1

SOME IMPORTANT POINTS

- Polynomial with one term called monomial
- Polynomial with two term called binomial
- Polynomial with three term called trinomial
- Polynomial of degree one is called linear polynomial
- Polynomial of degree two is called qluadratic polynomial
- Polynomial of degree three is called cubic polynomial
- The degree of a non-zero constant polynomial is zero
- A linear polynomial has only one zero
- A qluadratic polynomial has most two zero
- A cubic polynomial has most 3 zero
- Remainder theorem: let p(x) be any polynomial of degree is eater than or equal to one and let a be any real number if p(x) is divided by linear polynomial x-a then remainder is p(a).

- Factor these : x-a is a factor of the polynomial p(x) if p(a)=0. Also if x-a is a factor of p(x) then p(a)=0
- $(a+b)^2 = a^2+b^2+2ab$
- $(a-b)^2 = (a^2+b^2-2ab)$
- $(a^2-b^2) = (a+b)(a-b)$
- $(x + a)(x + b) = x^2+(a + b)x+ ab$
- $(a+b+c)^2 = a^2+b^2+c^2+2ab+2bc+2ca$
- $(a+b)^3 = a^3+b^3+3a^2b+3ab^2$
- $(a-b)^3 = a^3-b^3-3a^2b+3ab^2$
- $a^3+b^3+c^3 = (a+b+c)(a^2+b^2+c^2-ab-bc-ca)$
- $a^3+b^3 = (a+b)(a^2+b^2-ab)$
- $a^3-b^3 = (a-b)(a^2+b^2+ab)$

3. POLYNOMIALS 2

SOME IMPORTANT POINTS

- Polynomials having degree 1,2,3 are called quadratic cubic polynomials respectively
- Parabola cuts x-axis n times then number of zeros of polynomials is n.
- The standard form of quadratic polynomial is $ax^2+bx+c=0$. Where a, b and c are real numbers and a is not equal to 0.
- A quadratic polynomials can have at most 2 zeros and cubic 3.
- If α and β are zeros of the quadratic polynomials $ax^2+bx+c=0$, then $\alpha +\beta= -b/a$, $\alpha.\beta=c$
- If α, β and Υ are zeros of the quadratic polynomials $ax^2+bx+c=0$,

   then $\alpha + \beta + \Upsilon =-b/a$, $\alpha.\beta. \Upsilon=d/a$ and $\alpha.\beta+ \beta.\Upsilon+ \alpha.\Upsilon=-b/a$.
- The division algorithm is defined as If p(x) and g(x) are any two polynomials with g(x) not equal to 0 then we can find polynomials q(x) and r(x) such that
- $P(x)=q(x).g(x)+r(x)$ where r(x)=0 or degree of r(x) is less than degree of g(x)

3. POLYNOMIAL

1. Which of the following is not a polynomial?

   (a) x^3-x^2 (b) x^2-2x

   (c) x^2+2x (d) $x+1/x$

2. What is the value of x^0?

   (a) 0   (b) 1

   (c) x   (d) not defined

3. Which of the following is binomial?

   (a) $-5x^3$ (b) $5x+2x+3-x^2$

   (c) $-5x+2$ (d) $3+5x-x^2$

4. What is the degree of the polynomial $7y^6 - 7y^3 - 52y^8 - 7$?

(a) 6   (b) 3

(c) 7   (d) 8

5. The degree of polynomial $z^3 - z^4 - 2z + 7$

(a) z   (b) 3

(c) 4   (d) 7

6. Which of following is a linear polynomial?

(a) $4x^3 + 1$ (b) $4x^2 + 1$

(c) $4x + 1$ (d) $4x^2 + 4x - 1$

7. Which of the following is a polynomial in one variable?

(a) $7x^2 + 2y + 3$ (b) $3x^2 + 7x + 2$

(c) $9t^2 - 7t$ (d) all of these

8. that is the coefficient of a^2 $7a^3 - 3a^2 + 4a + 8$

(a) 7   (b) -3

(c) 3   (d) -7

9. What is the degree of the polynomial $7a^3 - 3a^2 + 8$?

(a) 7   (b) 8

(c) 3   (d) 9

10. $5x^2 - 7x + 4$ find the value of given polynomial if x=1

(a) -2   (b) 2

(c) 4   (d) 5

11. p (b) = $7b^2-7b+7$ find the value of b when b=3

    (a) 49   (b) 63

    (c) 70   (d) 56

12. P(x) = $9x^3-10x^2+2$ find the value of p(x) if x=-2

    (a) -30   (b) 30

    (c) 110   (d) -110

13. What is the zero of polynomial p(x) = 7x-7

    (a) -1   (b) 1

    (c) 0   (d) 7

14. What is the zero of polynomial p(x) = $3x^2+5x+2$

    (a) 1, 2/3   (b) -1, 2/3

    (c) -1, -2/3   (d) -2/3, 1

15. What is the zero of polynomial $7x^2-7x$

    (a) 0, 1   (b) 7,1

    (c) 0,-1   (d) -1, 2

16. Find the value of polynomial $3x^2-7x+2$ if x=-1

    (a) 6   (b) -2

    (c) -12   (d) 12

17. Find p (0) for polynomial x^2-x-1

    (a) 0   (b) -1

(c) 1   (d) 2

18.   Find p(2) for polynomial x^2+3x+5

       (a) -4   (b) 2

       (c) 3   (d) -5

19.   Find the zeroes of polynomial x^2+x-6

       (a) 2, -3   (b) 3,2

       (c) -2, -3   (d) -2, 3

20.   Find the zeroes of polynomial $4x^2-64$

       (a) -4, -4   (b) 4, 4

       (c) 4, 0   (d) 4,-4

21.   Find the zero of polynomial (7x-1) (x+2)

       (a) 1/7,-2   (b) -2, 7

       (c) 1, 7   (d) 2, 1/7

22.   Dividend =?

       (a) Divisor * remainder + quotient

       (b) Divisor*quotient + remainder

       (c) Quotient *remainder + divisor

       (d) None of these

23.   What is remainder when $3x^4-4x^3-3x-1$ is divided by x-1

       (a) -5   (b) 5

(c) 8   (d) 7

24.   What is the remainder if are divide $7x^2-3x+3$ by x-5

        (a) 170   (b) 162

        (c) 163   (d) 0

25.   What is the quotient obtained on dividing x^3+1 by x+1?

        (a) x^2-x+1 (b) x^2+x+1

        (c) $-x^2-x+1$ (d) $2x^2-x+2$

26.   What is the quotient obtained on dividing $8x^3-8$ by 2x-2?

        (a) $4x^2+4x-4$ (b) $4x^2-4x+4$

        (c) $4x^2-4x-4$ (d) none of these

27.   Find the remainder when $x^4-x^3+x^2-x+1$ divided by x-1

        (a) 1   (b) -1

        (c) 0   (d) -2

28.   Which is the multiple of polynomial x^2+x

        (a) x-2   (b) x+2

        (c) 2   (d) x-6

29.   What is the remainder when x^3-2x^2+2x+3 is divided by x-2

        (a) -7   (b) 8

        (c) 7   (d) 0

30.   What is the remainder when $3x^4-2x^3-5x+2$ is divided by

        (a) -6   (b) 12

        (c) 7   (d) 9

31. Which is the factor of x^3-343?

    (a) x-7   (b) x+7

    (c) x+8   (d) 8-x

32. Which of the following polynomial has factor x+2

    (a) x^3+4x^2+7x-6 (b) x^3+3x^2+5x+6

    (c) x^3+4x^2+5x-6 (d) x^3-18

33. Find the value of k if x-2 is factor $4x^4-3x^3+k$

    (a) -40   (b) 40

    (c) 38   (d) 4

34. What is the factor of polynomial x^2-x-6

    (a) (x+2) (x+3)   (b) (x-2) (x-3)

    (c) (x+2) (x-3)   (d) (x+3) (x-2)

35. What is the factor of polynomial $x^2-7x+12$?

    (a) (x+3) (x-4)   (b) (x-3) (x-4)

    (c) (x+3) (x+4)   (d) (x-3) (x+4)

36. (x+1) is a factor of polynomial

    (a) x^2+2x+1 (b) x^2-2x+1

    (c) $2x^2+5x-3$ (d) none of these

37. What is the value of $(x+y)^3 =$?

(a) $x^3+y^3+3x(x+y)$

(b) $x^3+y^3+3xy(x+y)$

(c) $x^3+y^3+3x^2y$

(d) $x^3+y^3+3xy^2$

38. Find the value of $(101)^3$

(a) 1030301

(b) 1020201

(c) 1010101

(d) 1001001

39. $x^3-y^3=$?

(a) $(x-y)(x^2-xy+y^2)$

(b) $(x-y)(x^2+xy+y^2)$

(c) $(x-y)(-x^2+xy+y^2)$

(d) $(x-y)(x^2-xy-y^2)$

40. What is the value of 104*105?

(a) 10520

(b) 10420

(c) 10920

(d) 10120

41. Find the value of k, if x-1 id factor $0/x^2-7x+k$

(a) 6

(b) 12

(c) 7

(d) 8

42. Find the value of k, if x+4 is factor 7x-4k

(a) 7

(b) -7

(c) 0

(d) 1

43. Find the value of k if x=0 is zero of $3x^2+-7x+k$

(a) 1

(b) 0

(c) 4 (d) 2

44. Find the polynomial whose zeroes are 4,-4

    (a) x^2-16 (b) x^2-10x-16

    (c) $3x^2$+16 (d) $4x^2$-4x-4

45. x^2-3x+2 find the value of polynomial of x=7

    (a) 51 (b) 21

    (c) 30 (d) -30

46. Find the product of (x-7) (x-6)

    (a) x^2-13x-42 (b) x^2-14x+42

    (c) x^2-13x+42 (d) none of these

47. What is the zero of polynomial x-7

    (a) 7 (b) -7

    (c) 0 (d) 1

48. What is the degree of polynomial $(3x^2$-7) (4x-4)+x?

    (a) 1 (b) 2

    (c) 3 (d) 4

49. What is the degree of polynomial 55?

    (a) 0 (b) 1

    (c) 2 (d) 55

50. What is the coefficient of x^2 in polynomial $7x^3$-$4x^2$-2x+8

    (a) 7 (b) -4

    (c) 4 (d) -2

QUE.	ANS.	QUE.	ANS.	QUE.	ANS.	QUE.	ANS.	QUE.	ANS.
1	D	11	A	21	A	31	A	41	A
2	B	12	D	22	B	32	B	42	B
3	C	13	B	23	A	33	A	43	B
4	D	14	C	24	C	34	C	44	A
5	C	15	A	25	A	35	B	45	C

6	C	16	D	26	B	36	A	46	C
7	A	17	B	27	A	37	B	47	A
8	B	18	C	28	A	38	A	48	C
9	C	19	A	29	C	39	B	49	B
10	B	20	D	30	B	40	C	50	B

4. QUADRATIC EQUATION

SOME IMPORTANT POINTS

- A quadratic equation is of the form of $ax^2+bx+c=0$ with a variable x
- A real number α is said to be the zero of the polynomial, when $a\alpha^2+b\alpha+c=0$
- We find the roots of the quadratic equation by factorised the quadratic equation in two linear factors and equating each factor to zero
- We can find the roots of the quadratic equation by quadratic formula
- $x= -b \pm \sqrt{b^{\wedge}2 - 4ac}$
- Where b^2-4ac is called the discriminate of the and it is $0 \leq b^2-4ac$
- A quadratic equation has two distinct real roots when $b^2-4ac > 0$
- A quadratic equation has two equal real roots when $b^2-4ac=0$
- A quadratic equation has no real roots when $b^2-4ac < 0$

4. QUADRATIC EQUATIONS

1. Find the discriminants if the roots of a quadratic equation are equal

   (a) 0   (b) 2

   (c) Less than zero   (d) 1

2. What is general form of a quadratic equation is (a $\neq$ 0).

   (a) ax +bx +cx=0   (b) ax+c

   (c) ax^2+ bx+c=0   (d) a+b+c=0

3. Which is a quadratic equation?

   (a) x^2+2= $(2+3)^2$ (b) x^2 +1/x=0

(c) x+1/x=2   (d) $x^2+bx+c=0$

4.   The roots of quadratic equation $x^2-3x+2=0$ are.

      (a) 1,2   (b) 1, -3

      (c) 1,-2   (d) -1, 2

5.   The quadratic equation $2x^2+kx+3=0$ have equal roots. Find the value of k.

      (a) $\pm 2\sqrt{3}$ (b) $\pm 2\sqrt{6}$

      (c) $\pm 8\sqrt{2}$ (d) $\pm 3\sqrt{2}$

6.   x+1/x=4 ¼, x $\neq$ 0 what is the roots

      (a) 2, ½   (b) 4, ¼

      (c) 4, ½   (d) 2,2

7.   If 1/x+2-1/x=3, x $\neq$ -2, x $\neq$ 0. Find the roots.

      (a) 3+$\sqrt{3}$ /3-$\sqrt{3}$, $\sqrt{3}$ -/3   (b) $\sqrt{3}$ -3/3, 3$\sqrt{3}$ /3

      (c) -3$\sqrt{3}$ /3, 3 $\sqrt{3}$ /3   (d) -3+ $\sqrt{3}$ /3, -3- $\sqrt{3}$ /3

8.   Find the quadratic equations $x^2-8x+16=0$.

      (a) 2,2   (b) 4,4

      (c) 1,1   (d) 0,4

9.   If roots are real. What is the nature of the roots of the quadratic equation: $5x^2-3x+2=0$?

      (a) roots are real   (a) 3, 2/$\sqrt{3}$

      (c) not real roots   (d) 5,3

10. If $3x^2+px+3=0$ has real roots. Find the value of p.

   (a) p $\geq$ 6 6 or p $\leq$ -6   (b) p $\leq$ 3 or p=3

   (c) p $\geq$ 3 or p $\leq$ -3   (d) p $\leq$ 2 or p $\geq$ 3

11. The product is 182 and sum is 27. Find the two numbers.

   (a) 13, 14   (b) 10, 20

   (c) 12, 11   (d) 18, 17

12. Find the quadratic equation.

   $2(2x+3/x-3)-2s(x-3/2x+3) =5$ x $\neq$ 3, x $\neq$ -3/2.

   (a) 4, 2   (b) 5, 4

   (c) 4, 1   (d) 6, 1

13. Find the solution of quadratic equation $3a^2+8abx+4b^2=0$; a $\neq$ 0

   (a) 3a/2, 2b/a   (b) 2b/a, 3a/2a

   (c) -2b/3a, 2b/a   (d) 4a/3b, 2a/3b

14. The one root of $3x^2=8x+(2k+1)$ is seven times the other, What is value of k.

   (a) -5/2   (b) 5/3

   (c) 4/3   (d) -4/3

15. If $ax^2-2\sqrt{5}x+4=0$ has equal roots. Find the value of a.

   (a) 4/5   (b) 5/4

   (c) 2/5   (d) -5/4

16. If roots are $4+\sqrt{7}/2$ and $4-\sqrt{7}/2$. Find the quadratic equation.

   (a) $4x^2-16x+9=0$ (b) $7x^2+4x-9=0$

   (c) $7x^2+4x+16=0$ (d) $4x^2+16x+9=0$

17. The equation $x^2-8x+p=0$ and roots are α and β and $\alpha^2+\beta^2=40$. Find the value of p.

    (a) 8   (b) 10

    (c) 14   (d) 12

18. Find the value of $\sqrt{\alpha}/\beta+\sqrt{\beta}/\alpha$. If $ax^2-bx+b=0$ and α and β are roots.

    (a) $\sqrt{b}/a$ (b) b/a

    (c) $\sqrt{a}/b$ (d) 2ab

19. If set of equation $4x^2-6x=0$ when $X \in N$.

    (a) $\emptyset$ (b) (2,1)

    (c) (1, 2)   (d) (0,1)

20. If $\sqrt{x^2}-16-(x-4)=\sqrt{x^2}-5x+4$. Find the set of equation.

    (a) (1,2)   (b) (4,5)

    (c) $(4,\sqrt{3})$ (d) none of these

21. Find the equation. Sum of roots is -1 and sum of their reciprocals is 1/6.

    (a) $x^3+x^2+6=0$ (b) $x^2+2x+6=0$

    (c) $x^2+x-6=0$ (d) none of these

22. If equation $x^2-bx/ax-c=k-1/k+1$ have roots are reciprocal. Find the value of k.

    (a) c+1/c-1   (b) c-1/c+1

    (c) c+1   (d) none of these

23. If $2x^2-5x-3=0$. Find the degree of equation.

   (a) 4   (b) 2

   (c) 5   (d) 3

24. Find the two roots of the equation $(x+4)(x+5) = 0$.

   (a) (2, 3)   (b) (-4, 5)

   (c) (4, -5)   (d) none of these

25. Solve the quadratic equation $y^2+2\sqrt{3}y+3=0$.

   (a) $(-\sqrt{3}, -\sqrt{3})$ (b) $(3, \sqrt{3})$

   (c) $(-\sqrt{3}, \sqrt{3})$ (d) $(-3, \sqrt{3})$

26. If equation $x+4/x=4$; $x \neq 0$ find the value of the x?

   (a) -2   (b) 4

   (c) 2   (d) -4

27. Find the quadratic equation whose one of the root is $3-\sqrt{5}$?

   (a) $x^2+2x+5=0$ (b) $x^2+3x+5=0$

   (c) $x^2-3x+5=0$ (d) $x^2-6x+4=0$

28. What is the value of the discriminant of the equation $\sqrt{3}x^2-2\sqrt{2}x-2\sqrt{3}=0$?

   (a) 39   (b) 32

   (c) 25   (d) 24

29. Divide 25 into two parts such that their product is 150?

   (a) 15 ,10   (b) 5 ,10

   (c) 20 ,10   (d) 30 ,20

30. If the quadratic equation is perfect square find the value of discriminant?

(a)1   (b)2

(c)0   (d)-1

31.   The product of Tara's age five years ago and his age nine years later is 15
       .What is the Tara present age?

       (a) 12 years   (b) 10 years

       (c) 6 years   (d) 8 years

32.   If $x^2+px+k=0$; is a quadratic equation and -4 is a root of equation and has
       equal roots what is the value of the k?

       (a) 9/4   (b) 3/5

       (c) 12/9   (d) 3/2

Answer:

Q	A	Q	A	Q	A	Q	A
1	A	9	C	17	D	25	A
2	C	10	A	18	A	26	A
3	C	11	A	19	A	27	D
4	A	12	D	20	B	28	B
5	B	13	C	21	C	29	A
6	B	14	A	22	A	30	C
7	D	15	B	23	D	31	C
8	B	16	A	24	B	32	A

NOTES

www.ingramcontent.com/pod-product-compliance
Lightning Source LLC
Chambersburg PA
CBHW080651180526
45168CB00008B/3380